YOUR KNOWLEDGE HAS VALUE

- We will publish your bachelor's and master's thesis, essays and papers

- Your own eBook and book - sold worldwide in all relevant shops

- Earn money with each sale

Upload your text at www.GRIN.com and publish for free

Bibliographic information published by the German National Library:

The German National Library lists this publication in the National Bibliography; detailed bibliographic data are available on the Internet at http://dnb.dnb.de .

Imprint:

Copyright © 2013 GRIN Verlag, Open Publishing GmbH
Print and binding: Books on Demand GmbH, Norderstedt Germany
ISBN: 9783668320246

Diego Issicaba

A Survey of Agent Technology Applications to Power Distribution Engineering

GRIN Publishing

A SURVEY OF AGENT TECHNOLOGY APPLICATIONS TO POWER DISTRIBUTION ENGINEERING

Prof. Diego Issicaba, PhD

2016

Contents

1. Introduction   2

2. Description of the Survey   2

3. Analysis and Discussions   3

4. Summary and Final Remarks   14

5. References   16

1 Introduction

Agent technology has been enlightened as the correct vector to promote decentralization, autonomous operation and active management activities in power distribution system operation. As a matter of fact, an adequate agent-based modeling can produce flexible, extensible, and robust systems, which are features of utmost importance to a smooth modernization of power systems. Moreover, the agent paradigm can provide a well-established notion of intelligence/smartness to be progressively applied along with the modernization of power distribution systems. Hence, in this document, an extensive survey about the applications of agent technology to power engineering is provided, aiming at helping engineers and academia to identify gaps in the state of the art to be explored in the future. This survey is presented and discussed highlighting the applications more directly related to the scope of the power distribution systems.

The document is organized as follows. In section 2, the developed survey is described, highlighting the size/range of the survey and main sources of information. Then, in section 3, discussions and analyzes about the state of the art are developed. Final remarks are outlined in section 4.

2 Description of the Survey

Recently, a representative literature survey on agent-based systems applied to power engineering was published [1] by the IEEE Power Engineering Society's Multi-Agent Systems Working Group, created in June 2005. The related publications were sought and categorized based upon their applications in protection, modeling, simulation, distributed control, monitoring and diagnoses. The survey included publications dated from 2001 to 2005 of the Intelligent Systems Applications to Power Systems (ISAP) conference proceedings as well as publications dated from 1998 to 2007 covering some IEEE transactions and some Institution of Electrical Engineers (IEE) journals. Hence, a similar survey was developed considering publications dated from 2001 to 2011 of the ISAP conference proceedings plus the publications dated from 1998 to 2011 of the IEEE Transactions on Power Systems, IEEE Transactions on Power Delivery, Institution of Engineering and Technology[1] (IET) Generation, Transmission & Distribution, Electric Power Systems Research (EPSR) and International Journal of Electrical Power & Energy Systems (IJEPES). Table 1 summarizes the outcomes of the survey emphasizing the

---

[1]In 2006, the IEE merged with the Institution of Incorporated Engineers to form the IET.

functional separation between power generation & transmission engineering and power distribution engineering[2]. Naturally, additional conference proceedings and journals were sought for the sake of completeness.

Table 1: Bibliographic survey of multi-agent systems applied to power engineering problems

	ISAP 2001–2003	ISAP 2005–2007	ISAP 2009–2011	IEEE/IEE IET Journals	EPSR/IJEPES Journals	Total
Gen. & Trans.	5 [2–6]	15 [7–21]	15 [22–36]	39 [1, 37–74]	3 [75–77]	77
Distr.	1 [78]	5 [79–83]	6 [84–89]	8 [90–97]	3 [98–100]	23
Total	6	20	21	47	6	100

Table 1 outlines in totality 100 publications, from which 23 were directly conducted aiming at power distribution system applications. In comparison with classical power engineering research fields such as reliability modeling, these figures suggest that the marriage between agent technologies and power engineering is far from being mature. This conclusion was expected given how agent-based systems have been maturing conceptually during the past few years, the interdisciplinary requirements to **building** practical applications under the agent paradigm, and the innovative character of applying such paradigm in power engineering solutions. On the other hand, although some overlapping work can be found, these figures also indicate that consistent research efforts have been made to exploit the agent paradigm to improve power engineering. These efforts cover from specific solutions proposed to improve the protection of small industrial systems [78] until abstract frameworks devised to manage entire bulk power systems [41].

3 Analysis and Discussions

Most probably, the application of agents to power system protection and primary control is the best starting point in what regards examining the evolution of academia in understanding the actual potential of agent technologies to power engineering. As highlighted in [1], Wooldridge's classical definition of an agent does not clearly distinguish agents from a number of existing softwares and hardware systems. Hence, under the quoted definition, controller devices or protective relays could be considered agents in the sense that they exhibit a certain degree of autonomy, they are situated in an environment (the power system), and they react to changes

---

[2]Differently from [1], papers related with the application of ant colony optimization algorithms as an approach for power system optimization problems were not included in the survey. The same applies to the direct application of reinforcement learning algorithms or other meta-heuristics.

(voltage and/or current signals) in the environment. The same reasoning applies, for instance, to an excitation system of a synchronous generator, a buck-boost transformer or a capacitor bank placed in a distribution feeder. Thus, following this crude definition, the term *agent* can be misapplied as a wrapping concept over control devices and protective relays, thereby originating multi-agent systems from interactions employed in legacy control and protective schemes. Although this can be arguably viewed as beneficial in providing different abstractions and ways of understanding existing systems, we share the harsh belief of some authors (e.g. [1]) which state that renaming existing or new systems built over existing technologies as *agents* offers almost nothing in terms of concrete engineering benefit.

Not surprisingly, one of the first research contributions [37] about agents to bulk power system protection applies a cooperation of agents abstracted over a wide range of equipments/components such as *line agents, bus agents, current transformer agents, potential transformer agents, circuit breaker agents, current transformer data collector agents,* and so forth, without an explicit representation of goal-directed behaviors for all of them. In power distribution systems, one example is the control and protection scheme described in [79] where a network split in zones, as proposed in [101], is managed by *monitoring agents, communication agents, breaker agents* and *relay agents,* which similarly lack goal-directed behavior. Also, in several further works, model validation through computer simulations (if any) did not imply that the agents were provided with data encapsulation, code encapsulation, separated threads of control, or the ability to run autonomously without being evoked externally (e.g. [97, 102]). Despite these issues, those works should not be seen with criticisms since they provided contributions to their respective fields. In fact, possible misconceptions might be merely faced as a consequence of the academic knowledge of the epoch in which the studies were performed.

More specifically to the applications of agent-based technology to power distribution system protection, the literature analysis had identified that the number of contributions/publications in this field of research is still limited. As one of the exceptions, we can refer to the multi-agent system applied to the overcurrent protection of a 6.6 kV industrial system proposed in [78]. The work focuses on using overcurrent *relay agents* and fuzzy sets [103] to search for an optimum protection capability scheme. Although agent simulation itself is not performed in sensu stricto, the idea of providing adaptive protection through agency, the cooperation among *relay agents,* and concerns regarding communication protocols are already discussed in the context of improving the protection schemes of power distribution systems. These concepts are present in more recent contributions such as [81, 104].

In [81], a coordinated protective scheme for distribution feeders is proposed using *feeder agents* and overcurrent *relay agents*. The scheme dictates that *relay agents* must update fault currents and maximum load currents using exchanged information whereas network topology changes are identified. For this accomplishment, source impedance information is fed downstream the primary substations while maximum loading information is fed upstream the loads. Object-oriented programming in C++ and agent communication language (ACL) are utilized. A small test system is used to illustrate the proposed rules and the resulting relay settings update seems promising, though more information could be exhibited about how the agents actually perform their reasoning. Also, it seems that more interactions and information would be necessary to achieve a practical usage of the scheme. For instance, the scheme does not seem directly extendable for distribution feeders with Distributed Generation (DG) units since time settings are not even mentioned.

In [104], the coordination of relays in power distribution systems with DG units using a multi-agent solution is proposed. Similarly to the previous works, it consists of applying the abstraction of a *relay agent*, but considering those along with *DG agents* and *equipment agents* to aid protective coordination. In summary, each *relay agent* takes sensory information, such as DG connection status, in order to choose among a list of settings. Then, some protection coordination is achieved through message exchange. The major drawback of the methodology is that agent message conveyance is required right after fault currents are identified. Besides demanding fast communications which are unfeasible to legacy systems, the idea of establishing agent interactions while the system devices are directly subjected to fault currents is, at least arguably, controversial. As a matter of fact, in case time constraints and backup protections were not adequately set, equipment life cycle might be jeopardized. Nevertheless, it is important to emphasize that some authors have been stating that while the cost of communications may not be justifiable to achieve sole protection functions, there will be situations where such cost might be justifiable if the same infrastructure is applied to provide other functions. This point of view is clearly exhibited, for instance, in the fault isolation scheme proposed in [97] where a wavelet-based fault direction identification technique is applied and *relay agents* interchange logical signals to locate the fault and activate circuit breakers.

Apart from devising protective schemes, agent-based systems have been applied to post-fault diagnosis as well. The first found work in the area refers to the ARCHON project [105, 106] whose main outcome is a general purpose architectural framework to facilitate cooperation among computational systems for industrial applications. As a practical deployment of the framework, the ARCHON approach was used to integrate reasoning systems within a control room including an *alarm analysis agent, breaker and relay supervision agent, black-out area identifier agent*, and even a *system restoration agent*, having as target the provision of useful information to the operators regarding contingency situations. Although some authors state that the resulting post-fault diagnosis application was not truly flexible or scalable [53], the general framework itself is considerably well thought in terms of deriving design concepts to structure interactions among problem solving entities (called agents in ARCHON's context) for industrial applications. Unfortunately, recent power engineering applications or even quotations about the ARCHON framework were not found in the power engineering literature, a fact that might be explained by the lack of dissemination of ARCHON framework's features in power engineering journals and conference proceedings.

Almost ten years ahead, the protection engineering diagnostic agents (PEDA) proposed in [53, 63, 107] were introduced. In this research, a post-fault diagnose assistant system to protection engineers was developed through the integration of several protection analysis tools. The integration was performed by wrapping analysis tools in agents named *protection validation and diagnoses agent*, *incident and event identification agent*, *fault record interpretation agent*, *fault record retrieval agent*, *collation agent* and *engineering assistant agent*. Despite of the straightforward (but well justified) aspect in how the authors defined the agent types, the work stands as one of the few where some multi-agent system design concepts were applied, namely the task decomposition stage of the DESIRE [108] methodology. Codding was performed using ZEUS [109] and at least one agent descriptor with functional task and exchange resource information is provided, what we believe should be a practice to be adopted in exhibiting research material in this area. The work itself is discussed in the context of bulk power system applications. Nevertheless, with the ongoing increase of feeder automation, it could be adapted/extended to distribution feeder applications. Other related system that could be exploited on distribution feeder automation is the conditional monitoring multi-agent system (COMMAS) currently applied to plant monitoring [110] and transformer monitoring [28, 57].

Regarding power distribution system restoration, we initially instantiate the multi-agent system introduced in [111, 112]. This work resembles a previous research developed by the same authors in the context of bulk power system restoration [49] and consists of abstracting *feeder agents* and *load agents* to perform restoration functions according to the following directives.

1. Once a service interruption is assigned, load agents have to isolate their respective loads and send a message to their service feeder agents.

2. The feeder agents receive the messages from the load agents and handle the restoration using rules extracted from operator's experience.

The proposal has the merit of being one of the earliest in the area. However, decision-making about the restoration procedures is still centralized on the feeder agents while the role of the load agents is basically to notify the feeder agent about service interruptions. In addition, discussions about software modeling and the practicality of implementing one agent per secondary transformer in actual power distribution systems are neglected. Decisions about restoration are also centralized in the solutions proposed in [86, 102], where *global agents* [102] and *restoration leader agents* [86] were envisioned to wrap reconfiguration and decision-tree algorithms, respectively. The latter solution [86] is assumed to follow the BDI agent architecture but beliefs, desires and intentions are not explicitly described in the agent restoration mechanisms.

A decentralized multi-agent solution for power distribution system restoration was proposed in [96] where three agent types are abstracted: the *generator agent* (to model primary substation sources), the *load agent* and the *switch agent*. Service restoration is then achieved through interactions among these entities without the figure of a higher level entity managing the switch actions and taking into account the available transfer capacity of switches. The work is clearly JADE-oriented and the agents are described according to their initial knowledge and behaviors. Also, the simple UML class diagram of the application is presented making easy understanding the author's implementation. Further improvements are introduced in [113] with a small update to consider rules for load shedding and a DG agent. Although engineering aspects about the actual implementation of the proposed solution are not discussed, these publications succeed on describing simple behaviors and rules to aid the restoration processes in power distribution systems. Furthermore, the provided small case study is well described, so that we recommend its implementation to power system researchers interested in achieving some proficiency on implementing agent behaviors in JADE.

In concise terms, the contributions in [86, 96, 102, 111–113] would seem more realistic in case, for instance, the load agents explicitly represented an aggregation of customers, as well as requirements for monitoring and main hypothesis about infrastructure were clarified. Conversely, other contributions such as in [91–94, 114] introduce fair hypotheses also offering discussions about the practicality of their proposed solutions. As a matter of fact, the authors in [91] present an interesting application of agency to condition assessment and fault management, though much more focused on engineering solutions for power distribution system automation than in agent modeling itself. The concept is composed of three main aspects as follows.

1. A software object for secondary substations which encapsulates a set of object classes representing the substation hardware and following the concept of logical nodes from the IEC 61850 [115].

2. Functionalities based on token (permission to act and execute local functions) message conveyance between neighboring substations. The basic procedure implies that, when a substation receives a token, it executes the required function, attaches the result, and conveys information to downstream substations. After processed at the last secondary substation, the token is sent back to the primary substation.

3. An information access model which, in summary, hierarchically defines that a permission to execute a given function is conveyed from the control center to the primary substation, and further downstream the secondary substations.

Hence, fault location and isolation are obtained by sending a token from a primary substation downstream secondary substations to verify fault indicators and to open normally closed switches around the faulted section. Power is restored by closing the circuit breaker at the primary substation and passing tokens upstream towards alternative supply primary substations. The proposed concept was extended to include state estimation in [92] and has a reduced engineering complexity in the sense that all secondary substations in a feeder are copies of a common secondary substation type. Furthermore, it provides local management at the secondary substation level as well as it reduces the information and communication saturation since, for the developed functions, the control centers "see" a primary substation area as a single entity. Regarding communication issues and validation tests, the authors verified the reduction in number of hops for fault management in comparison with a totally centralized control approach, and they tested the approach using a small but very illustrative prototype implementation.

Another agent-based approach much more focused on finding engineering solutions to power distribution system automation than in agent modeling itself is provided in [114]. The authors introduce an hierarchical automation architecture based on the concept of *intelligent logical nodes*, which is envisioned as an extension of the logical node (see IEC 61850 [115]) concept applied to substation automation. Although the rules for fault location and power restoration can be found in previous works, the approach itself is very interesting in the sense that it establishes a direct integration of the IEC 61850 and IEC 61499 [116] into the so-called intelligent logical nodes. Concisely, the integration dictates that for each logical node, an intelligent logical node is implemented as a composite function block of the IEC 61499 with a database, service interpreter and intelligence (the part responsible for decision making and negotiation). The agents communicate via services of the IEC 61850 and computer validation tests were carried out using a MATLAB [117] model interfacing with function block models through custom-design user datagram protocol sockets.

Following a different approach, a multi-agent system for fault location, isolation and power restoration is described in [93]. The approach is directed to underground power distribution systems and assumes three software agent types located at the secondary distribution substations. They are the *agent expert* that must handle emerging situations, the *agent inter* that provides connection with the physical environment, and *agent com* that is responsible for communications. Also, a *terminal agent* performs some activities at the primary distribution substations. All these agents have specific rules to be applied in three operation states called: steady-state, fault isolation state and restoration state. The research has continued in [94] where the engineering complexity [91] was reduced in considering only two agent types: the *primary substation agent* and the *secondary substation agent*. Hence, the resulting structure resembled the one proposed in [91,92], though fault isolation and power restoration are achieved differently by using some JADE-oriented implementations of agent behaviors. The applications in [94] are sound and the rules behind the agent behaviors consolidate the idea of employing neighborhood interactions as in [91,96]. The economic feasibility of these solutions were verified in [99].

In order to close (for now) the discussions about the agent-based solutions to power distribution system restoration, we quote the actual application of the IntelliTEAM II automatic restoration system developed by the S&C Electric Company [118]. In the approach, the basic unit of operation is a line segment bounded by *intelligent switching points*. The combination of the line segment and intelligent switching points is called a *team* and a so-called virtual agent manages internal team information sharing and restorative tasks. Also, a *contract agent* that works across multiple teams back to a common load source is applied to avoid safety problems and a *return to normal agent* is utilized to handle backing to normal configuration after a network failure is repaired. Current implementations usually rely on peer-to-peer wide area communications provided by a 900 MHz UtiliNet spread spectrum radio system that transports data using a connection-less mesh architecture, thus allowing each radio in the network to act both as a repeated radio and a data transceiver for an automated switch. Even though agency is not dealt specifically in the abstraction (e.g. by recurring on a reasoning model), it is a fact that the company have been installing the automation solutions since 2003 with success, assigning both interest and economic feasibility to providing distributed restorative functions for power distribution systems. Applications in the field have reported [119] automatic restoration cases which endured less than 10 seconds, which is an impressive result.

Up to this point, we have outlined agent-based applications specifically leaned to power distribution system protection, diagnoses, fault isolation, fault location and power restoration. However, other activities such as voltage control, power flow management and market operation were also approached under the agent paradigm either in a sole way or within broad conceptual frameworks. As one example of the latter, let us take the "cell" concept introduced in [120], where a network cell is viewed as a self-managing entity of protection, voltage control and power control activities. Summarily, the concept produces a system structure which is envisioned to evolve in analogy with a biological cell division process from a whole distribution network as a single cell to a final distribution network probably composed of thousands of cells. Hence, the resulting theoretical framework was recently utilized up to a certain degree in [121] where an agent-based converter interface for active distribution networks is proposed. The converter was named *smart power router* and it was applied to interfacing different cells of a power distribution system. The work uses an operation structure quite similar to the proposed in [122], where each agent is considered an autonomous actor handling the three issues: management (performs the object functions for voltage regulation, power management, or state estimation), coordination (defines control set points for the object functions) and execution (activates control actions). Aspects of reasoning modeling were not derived and the focus was strictly on the control functions which were further improved with optimal power flow computations [123]. Nevertheless, interesting results were provided and the solutions were verified under a laboratory setup specifically configured for the application.

One of the most important aspect of analyzing the introduction of the cell concept in [120] is to verifying that utilities have the pragmatic (and reasonable) perception that it is impossible to transform a distribution network in an active distribution network overnight. It is therefore necessary to abstract the current situation so that the evolution to active distribution networks happens in a controlled manner aiding network designers to plan their modifications and extensions in alignment with a long-term vision [120]. These concerns were also raised in developing the autonomous regional active network management system (AuRA-NMS) [82], though the resulting solutions were centralized in primary substation equipments and mostly tailored for regional-scale applications. Concisely, the AuRA-NMS is composed of functionalities wrapped in agents, namely the *power flow management agent, voltage control agent,* and *automatic restoration agent.* Also, an additional agent named *arbitration agent* was envisioned to avoid conflicting actions by imposing priorities, though more information could be given about its reasoning. These agents were deployed within an agent platform running across several ABB COM6xx computers [124] placed at primary substations. The functionalities were approached as follows.

1. Power flow management: A constraint satisfaction problem formulation was applied to power flow management [88] where DG control signals (100%, 90%, 80%, and so forth, of rated power) represent the domain of discrete variable values. Hence, potential solutions to maximize DG access are checked while power flow and contractual (e.g. "last-in, first-off") limits impose the constraints to the problem. A solver is utilized to find multiple ranked solutions to a given problem supporting operator's requirements for graceful degradation. Thus, if the preferred solution does not mitigate thermal excursions due to model error or measurement error, the next ranked solution can be implemented. Optimal power flow arrangements were under trials as well but concerns regarding algorithm non-convergence seem to deviate the authors from this solution. They state (and we agree) that from operator's perspective optimality plays second fiddle to robustness, thus suboptimal but robust solutions are preferable instead of those which strive for optimality.

2. Voltage control: A case-based reasoning approach was used [98] to avoid problems associated with the non-convergence of power flows or optimal power flows. The idea was to apply a set of preenumerated solutions created over a representative set of voltage excursions aiming at providing solutions to similar but different voltage excursions. The authors considered as control measures the change of tap positions, DG power factor set points and DG active power outputs. A constraint programming arrangement was also considered but issues regarding to search space size and computation time seems to drive the authors towards the case-based reasoning approach as main solution, though such statement was never explicitly mentioned in [88].

3. Automatic restoration: Researchers at the University of Cardiff developed a reconfiguration approach to minimize the number of disconnected customers based on the knowledge of fault location and pre-fault loadings [125]. No further information was found about this functionality up to April 2012.

The work provides interesting discussions about operator's requirements for active management of distribution systems and manifests explicit goal-driven concerns in the solutions design, mostly in the sense a control engineer would be able to set control goals for an area under the AuRA-NMS. The research project was well planned and executed producing further contributions and extensions [88, 98, 125, 126].

Note that all these agent-based solutions and active management applications did not rely on a market operator to devise real-time operation, which is reasonable since the great majority of the customers connected to the power distribution systems pay a pre-specified tariff for the electric energy without any sort of real-time selling/buying activity. As a matter of fact, achieving operation through a market interactive customer per household seems unfeasible in the short/mid-term given that even different payments for different levels of power quality are far from being popular. Despite of these issues, some authors have also studied market operations to power distribution systems under the agent paradigm. In this context, the market-based control concept to supply-demand matching in electricity networks [127], called PowerMatcher, must be quoted. The objective of the concept is the optimal use of devices (named agents) which sell/buy electricity on a market exchange and are categorized as stochastic operation, shiftable operation, external resource buffering, electricity storage and user-action. The resulting application is sound and field test implementations are shown in [128], though the authors are not clear about how agent-based simulations were actually performed.

A market operation control concept is also explored in a micro grid concept in [129] where the agents are assigned to a power system as well as to micro grid entities such as the micro grid central controller, the production units and the consumption units. This work has continued in [18, 19, 85, 90, 130–132] where JADE-oriented behaviors, technical discussions and test cases are presented. In architectural terms, the main difference from the multi-micro grid paradigm lies in the fact that, in these works, the micro grid communicates directly to a technical operator and a market operator, while in the multi-micro grid paradigm proposed in [133] the micro grids interact with a controller placed nearby the primary substation. The multi-agent system developed for resource schedule in multi-micro grids shown in [100] also shares such a difference.

Another set of market-based solutions over a broad framework is obtained through the Infotility GridAgents [134] software platform deployed at the Consolidated Edison Company of New York's power distribution system. In summary, the GridAgents were initially designed by Infotility for energy transactions based on real time pricing signals. Recently, the GridAgents framework sells a set of default decision support solutions to power systems split in the activities of Resource (pull information and write data from meters, EMS, sensors, databases), Interface (display modifications), Task (EMS functions) and Broker (market/contract operator). For distribution network control, the developments are a set of plug-in modules with embedded solutions such as to compute optimal DER response to price signals, inclusion of business rules, physical constraints, and artificial intelligence learning routines to optimize operation and/or response over time. The most interesting feature of the framework is the integration with internet-based systems through a web services gateway via blackboard agents.

Finally, the literature analysis also showed that the design of voltage control schemes for distribution feeders using agent-based technology is a recent and rare research topic to be found.

In a set of exceptions, we initially quote the broad hierarchical control framework for transmission and distribution networks described in [135]. This framework employs a hierarchical control architecture in which actions follow a chain of command from a top layer (control center) to bottom layers (power distribution systems and loads). As consequence, interactions are performed by a central *EMS agent* and layered *relay agents*, all implemented in JADE. The key aspect of the approach is a reactive load control optimization algorithm to improve voltage profiles in power distribution systems. However, a *top feeder relay* attributed to each distribution feeder is envisioned to solve the optimization problem and to coordinate bottom controllers, which in turn simply corresponds to delegating centralized voltage coordination decisions to an entity at the primary substation level instead of relying on functions (or operators) at a control center. This sort of delegation was also exhibited in the previously described voltage control of AuRA-NMS [98], but the latter approach is much more connected to the actual practice of operation mainly in the sense it focuses on pre-enumerated solutions which can be thoroughly analyzed by operators and experts instead of relying on an algorithm designed to strive for optimality.

The straightforward delegation of voltage coordination functions is avoided in [136] where Remote Terminal Units (RTUs) are placed at DG sites and near capacitors to aid control by the voltage regulators. The harsh drawback of the work is that the authors stated in the publication abstract that the RTUs coordinated together compose a multi-agent system, but actually agent modeling and interactions are not described in the contents of the document. Conversely, agent interactions through the contract net protocol are explicitly applied in the multi-agent reactive power dispatch of DG units proposed in [95]. Nevertheless, the authors did not considered utility voltage regulators or shunt capacitors and they did not revealed a proper simulation model to validate their work, as also identified in [137]. At last, we refer to the well described coordination mechanisms for voltage control introduced in [137] which considers the following agents: *load tap changer control agent, voltage regulator control agent, DG control agent* and *shunt capacitor control agent*. In this approach, each agent type has an internal architecture where control rules/equations are wrapped into modules named perception, interpretation, expert-based decision-maker, and execution. Decision-making is performed using expert-based pre-specified lookup tabled rules whereas *load tap changer control agents* and *voltage regulator control agents* receive information from the other agents regarding updated and forecasted set points, as well as solution proposals (or requests) enforced due to voltage violations which can (or cannot) be solved locally. All agents are implemented in JADE whilst sensors and actuators are modeled in MATLAB [117] Simulink. The work stands out in the sense that the objectives

of the agents (minimize voltage deviations, injected reactive power and tap/switch operations)
are explicitly defined by the authors, which combined with the module design indicates some
goal-oriented behavior is considered in the modeling and implementation.

4   Summary and Final Remarks

The extensive survey about the applications of agent systems to power engineering brought
up several insights and conclusions regarding how power engineers have been exploiting agent
technologies to their end means. As previously stated, in comparison with classical power
engineering research fields, such exploitation is far from being mature due to, for instance, the
interdisciplinary requirements to build practical applications under the agent paradigm and the
innovative character of applying such paradigm in power engineering solutions. The existing
applications are fragmentally distributed among several research areas and, differently from
what we expected in the beginning of the survey, the research activities have not been booming
recently, even with the rise of smart/modern grid trends.

Applications to power distribution engineering are even more fragmentally distributed. Also,
common ground is missing in what regards how solutions in protection, monitoring, diagnoses,
voltage control, power flow management, fault location, fault isolation and power restoration
can be integrated to support Distribution Management System functions. As one of the main
conclusions of the survey, we identified that

> *there is a lack of an agent-based architecture specifically designed to support dis-
> tributed feeder applications aligned to the trends enforced by the smart grid paradigm.*

Without such an architecture, integrating solutions from different areas that influence distribution feeder operations can become a task full of complexities. This point is particularly
interesting since one of the main justifications for applying agent systems lies in designing extendable solutions with reduced complexity. In fact, several of the surveyed works are quite
extendable mostly in the sense of accommodating new agent instances of a designed agent type.
However, it is not a straightforward task to accommodate interactions of agent types developed under different contexts. As an example, let us consider the voltage control mechanisms
in [137] and the IntelliTEAM II [118] restoration mechanisms. It would be advisable to avoid
the change of device tap positions at the same time restoration switch procedures are enforced
and this could be obtained, for instance, by prioritizing goals. Nevertheless, since goal-directed
behavior and agent planning are not explicit in IntelliTEAM II, one cannot straightforwardly

choose which *team* plans to disable in order to prioritize a goal in [137]. With this simple example, we illustrate that

> *the lack of explicit representation of goal-directed behaviors interrelated with agent planning makes difficult to conjugate in a common framework the agent solutions proposed in the literature.*

As a matter of fact, the great majority of works which perform some sort of validation through computer simulations applies the middleware JADE to aid implementation. The JADE platform focuses on implementing the FIPA reference model providing communication infrastructure, platform services such as agent management, and a set of development and debugging tools. On the other hand, it intentionally leaves open much of the issues of internal agent concept, which in turn can be considered through add-ons as in JADEX [138]. Hence, in order to develop agent systems, it is necessary to consider the intra-agent as well as inter-agent structures, being the BDI model one of the main options to enable viewing an agent as a goal-directed entity that acts in a rational manner. Nevertheless, the only application of BDI models in the power distribution system environment is found in [86], and even in this work beliefs, desires and intentions are not properly described. As conclusion, it is crucial that

> *some agent architectural models, such as the BDI model, are included in the agent systems developments, thereby avoiding the sole use of the middleware JADE by exploring other alternatives such as JADEX or JASON.*

Another issue that jeopardizes the possibility of conjugating agent developments in power distribution engineering is the lack of deployment of designing tools and software engineering methodologies. The surveyed works do not describe their designing, with the few exceptions of the very closed applications of the ARCHON framework and the PEDA agents. Usually, agent types are chosen according to the convenience of the designer without proper clarifications. Once design matters are not described, the design itself cannot be extended directly which is a frustrating dichotomy for those interested in customizing a set of distributed solutions to their particular applications. Therefore,

> *power engineering academia should be aware that the description of software engineering design matters are of utmost importance to guarantee their proposed agent systems can be actually extended in different contexts and applications. Also, the usage of well established and documented agent design methodologies, such as Prometheus,*

is appreciated to foster proper discussions about the virtues and drawbacks of each development.

Finally, none of the surveyed works provides one of the most important issue for the practical implementation and acceptance of agent-based technology in power distribution engineering. In order to justify altering the power distribution system infrastructures and providing alternative distributed functionalities

it is necessary an environment model which emulates the system operation to evaluate the long-term impact of the agent-based solutions according to standardized (and regulated) power distribution system performance indices.

The absence of such environment simulation is understandable since this would require a simulation model which considers altogether aspects from power distribution system planning and infrastructure, the long-term failure/repair cycle of the system components, topology changes caused by protective and control network actions, DER impact on system steady-state and dynamics, control strategies such as for islanding and load shedding, and so forth.

All these issues can be exploited to improve the applications of agent technology to power engineering solutions.

5 References

[1] S. D. J. McArthur, E. M. Davidson, V. M. Catterson, A. L. Dimeas, N. D. Hatziargyriou, F. Ponci, and T. Funabashi. Multi-agent systems for power engineering applications – Part I: Concepts, approaches, and technical challenges. *IEEE Transactions on Power Systems*, 22(4):1743–1752, Nov. 2007.

[2] E. E. Mangina, S. D. J. McArthur, and J. R. McDonald. Multi agent system knowledge representation for power plant condition monitoring based on modal logic. In *Proceedings of the International Conference on Intelligent System Applications to Power Systems (ISAP)*, Budapest, Hungary, Jun. 2001.

[3] F. R. Monclar and R. Quatrain. Simulation of electricity markets: A multi-agent approach. In *Proceedings of the International Conference on Intelligent System Applications to Power Systems (ISAP)*, Jun. 2001.

[4] M. A. Sanz-Bobi, J. Villar, E. F. Sánchez-Úbeda, L. F. Robledano G. Plaza, S. Revuelta, and A. Kazi. Diamond: Multi-agent environment for intelligent diagnosis in power systems. In

Proceedings of the International Conference on Intelligent System Applications to Power Systems (ISAP), Budapest, Hungary, Jun. 2001.

[5] R. Gao and L. H. Tsoukalas. Anticipatory paradigm for modern power system protection. In *Proceedings of the International Conference on Intelligent System Applications to Power Systems (ISAP)*, Lemnos, Greece, Aug.–Sep. 2003.

[6] N. Yamaguchi, N. Watanabe, K. Okada, I. Watanabe, M. Nagata, and H. Asano. Development of a negotiation process model by agent system for electricity bilateral contracts. In *Proceedings of the International Conference on Intelligent System Applications to Power Systems (ISAP)*, Lemnos, Greece, Aug.–Sep. 2003.

[7] J. S. Heo and K. W. Lee. A multi-agent system-based intelligent steady-state model for a power plant. In *Proceedings of the International Conference on Intelligent System Applications to Power Systems (ISAP)*, Arlington, Virginia, USA, Nov. 2005.

[8] T. Ono and G. C. Verghese. Replicator agents for electricity markets. In *Proceedings of the International Conference on Intelligent System Applications to Power Systems (ISAP)*, Arlington, Virginia, USA, Nov. 2005.

[9] N. W. Oo and V. Miranda. Evolving agents in a market simulation platform – a test for distinct meta-heuristics. In *Proceedings of the International Conference on Intelligent System Applications to Power Systems (ISAP)*, Arlington, Virginia, USA, Nov. 2005.

[10] N. Yu, C-C. Liu, and L. Tesfatsion. Modeling of suppliers' learning behaviors in an electricity market environment. In *Proceedings of the International Conference on Intelligent System Applications to Power Systems (ISAP)*, Niigata, Japan, Nov. 2007.

[11] K. Huang, D. A. Cartes, and S. K. Srivastava. A multi-agent based algorithm for mesh-structured shipboard power system reconfiguration. In *Proceedings of the International Conference on Intelligent System Applications to Power Systems (ISAP)*, Arlington, Virginia, USA, Nov. 2005.

[12] T. Hiyama and T. Ohta. Hierarchical stabilization control with upper level multi-agent based controller and lower level local controller. In *Proceedings of the International Conference on Intelligent System Applications to Power Systems (ISAP)*, Arlington, Virginia, USA, Nov. 2005.

[13] J. M. Solanki and N. N. Schulz. Using intelligent multi-agent systems for shipboard power systems reconfiguration. In *Proceedings of the International Conference on Intelligent System Applications to Power Systems (ISAP)*, Arlington, Virginia, USA, Nov. 2005.

[14] H. Qi, W. Zhang, and L. M. Tolbert. A resilient real-time agent-based system for a reconfigurable power grid. In *Proceedings of the International Conference on Intelligent System Applications to Power Systems (ISAP)*, Arlington, Virginia, USA, Nov. 2005.

[15] V. M. Catterson, E. M. Davidson, and S. D. J. McArthur. Issues in integrating existing multi-agent systems for power engineering applications. In *Proceedings of the International Conference on Intelligent System Applications to Power Systems (ISAP)*, Arlington, Virginia, USA, Nov. 2005.

[16] T. Nagata, H. Fujita, and H. Sasaki. Decentralized approach to normal operations for power system network. In *Proceedings of the International Conference on Intelligent System Applications to Power Systems (ISAP)*, Arlington, Virginia, USA, Nov. 2005.

[17] A. C. Tellidou and A. G. Bakirtzis. Agent-based analysis of monopoly power in electricity markets. In *Proceedings of the International Conference on Intelligent System Applications to Power Systems (ISAP)*, Niigata, Japan, Nov. 2007.

[18] A. L. Dimeas and N. D. Hatziargyriou. Agent based control of virtual power plants. In *Proceedings of the International Conference on Intelligent System Applications to Power Systems (ISAP)*, Niigata, Japan, Nov. 2007.

[19] A. L. Dimeas and N. D. Hatziargyriou. Design of a mas for an island system. In *Proceedings of the International Conference on Intelligent System Applications to Power Systems (ISAP)*, Niigata, Japan, Nov. 2007.

[20] K. Huang, S. K. Srivastava, D. A. Cartes, and M. Sloderbeck. Intelligent agents applied to reconfiguration of mesh structured power systems. In *Proceedings of the International Conference on Intelligent System Applications to Power Systems (ISAP)*, Niigata, Japan, Nov. 2007.

[21] O. Gehrke and H. Bindner. Building a test platform for agents in power system control: Experience from syslab. In *Proceedings of the International Conference on Intelligent System Applications to Power Systems (ISAP)*, Niigata, Japan, Nov. 2007.

[22] E. Ciapessoni and E. Corsetti. Defplans: Agent modeling techniques for power system emergency control. In *Proceedings of the International Conference on Intelligent System Applications to Power Systems (ISAP)*, Curitiba, Brazil, Nov. 2009.

[23] J. H. van Sickel and K. Y. Lee. Distributed discrete event and pseudo real-time combined simulation for multi-agent controlled power plants. In *Proceedings of the International Conference on Intelligent System Applications to Power Systems (ISAP)*, Curitiba, Brazil, Nov. 2009.

[24] J. A. Momoh, K. Alfred, and Y. Xia. Framework for multi-agent system (MAS) detection and control of arcing of shipboard electric power systems. In *Proceedings of the International Conference on Intelligent System Applications to Power Systems (ISAP)*, Curitiba, Brazil, Nov. 2009.

[25] P. C. Baker, V. M. Catterson, and S. D. J. McArthur. Integrating an agent-based wireless sensor network within an existing multi-agent condition monitoring system. In *Proceedings of the International Conference on Intelligent System Applications to Power Systems (ISAP)*, Curitiba, Brazil, Nov. 2009.

[26] M. A. Rosa, A. M. Leite da Silva, V. Miranda, M. Matos, and G. Sheblé. Intelligent agent-based environment to coordinate maintenance schedule discussions. In *Proceedings of the International Conference on Intelligent System Applications to Power Systems (ISAP)*, Curitiba, Brazil, Nov. 2009.

[27] P. Oliveira, T. Pinto, H. Morais, Z. A. Vale, and I. Praca. MASCEM – an electricity market simulator providing coalition support for virtual power players. In *Proceedings of the International Conference on Intelligent System Applications to Power Systems (ISAP)*, Curitiba, Brazil, Nov. 2009.

[28] V. M. Catterson, S. E. Rudd, S. D. J. McArthur, and G. Moss. On-line transformer condition monitoring through diagnostics and anomaly detection. In *Proceedings of the International Conference on Intelligent System Applications to Power Systems (ISAP)*, Curitiba, Brazil, Nov. 2009.

[29] U. Hager, S. Lehnhoff, C. Rehtanz, and H. F. Wedde. Multi-agent system for coordinated control of facts devices. In *Proceedings of the International Conference on Intelligent System Applications to Power Systems (ISAP)*, Curitiba, Brazil, Nov. 2009.

[30] J. H. van Sickel and K. Y. Lee. Real-time based agent architecture for power plant control. In *Proceedings of the International Conference on Intelligent System Applications to Power Systems (ISAP)*, Curitiba, Brazil, Nov. 2009.

[31] T. Pinto, Z. Vale, F. Rodrigues, I. Praca, and H. Morais. Cost dependent strategy for electricity markets bidding based on adaptive reinforcement learning. In *Proceedings of the International Conference on Intelligent System Applications to Power Systems (ISAP)*, Athens, Greece, Sep. 2011.

[32] A. F. Shosha, P. Gladyshev, S-S. Wu, and C-C. Liu. Detecting cyber intrusions in SCADA networks using multi-agent collaboration. In *Proceedings of the International Conference on Intelligent System Applications to Power Systems (ISAP)*, Athens, Greece, Sep. 2011.

[33] D. Papadaskalopoulos, P. Mancarella, and G. Strbac. Decentralized, agent-mediated participation of flexible thermal loads in electricity markets. In *Proceedings of the International Conference on Intelligent System Applications to Power Systems (ISAP)*, Athens, Greece, Sep 2011.

[34] T. Nagata and S. Fukunaga. An autonomous distributed agent approach to power system restoration. In *Proceedings of the International Conference on Intelligent System Applications to Power Systems (ISAP)*, Athens, Greece, Sep. 2011.

[35] A. Saleem, L. Nordstrom, and M. Lind. Knowledge based support for real time application of multiagent control and automation in electric power systems. In *Proceedings of the International Conference on Intelligent System Applications to Power Systems (ISAP)*, Athens, Greece, Sep. 2011.

[36] D. Divenyi and A. Dan. Simulation results of cogeneration units as system reserve power source using multiagent modeling. In *Proceedings of the International Conference on Intelligent System Applications to Power Systems (ISAP)*, Athens, Greece, Sep. 2011.

[37] Y. Tomita, C. Fukui, H. Kudo, J. Koda, and K. Yabe. A cooperative protection system with an agent model. *IEEE Transactions on Power Delivery*, 13(4):1060–1066, Oct. 1998.

[38] V. Krishna and V. C. Ramesh. Intelligent agents for negotiations in market games. Part I: Model. *IEEE Transactions on Power Systems*, 13(3):1103–1108, Aug. 1998.

[39] V. Krishna and V. C. Ramesh. Intelligent agents for negotiations in market games. Part II: Application. *IEEE Transactions on Power Systems*, 13(3):1109–1114, Aug. 1998.

[40] C. S. K. Yeung, A. S. Y. Poon, and F. F. Wu. Game theoretical multi-agent modelling of coalition formation for multilateral trades. *IEEE Transactions on Power Systems*, 14(3):929–934, Aug. 1999.

[41] M. Amin. Toward self-healing energy infrastructure systems. *IEEE Computer Applications in Power*, 14(1):20–28, Jan. 2001.

[42] E. E. Mangina, S. D. J. McArthur, J. R. McDonald, and A. Moyes. A multi agent system for monitoring industrial gas turbine start-up sequences. *IEEE Transactions on Power Systems*, 16(3):396–401, Aug. 2001.

[43] H. F. Wang. Multi-agent co-ordination for the secondary voltage control in power-system contingencies. *IEE Proceedings – Generation, Transmission and Distribution,*, 148(1):61–66, Jan. 2001.

[44] E. E. Mangina, S. D. J. McArthur, and J. R. McDonald. Reasoning with modal logic for power plant condition monitoring. *IEEE Power Engineering Review*, 21(7):58–59, Jul. 2001.

[45] D. W. Bunn and F. S. Oliveira. Agent-based simulation-an application to the new electricity trading arrangements of england and wales. *IEEE Transactions on Evolutionary Computation*, 5(5):493–503, Oct. 2001.

[46] P. Wei, Y. Yan, Y. Ni, J. Yen, and F. E. Wu. A decentralized approach for optimal wholesale cross-border trade planning using multi-agent technology. *IEEE Transactions on Power Systems*, 16(4):833–838, Nov. 2001.

[47] D. V. Coury, J. S. Thorp, K. M. Hopkinson, and K. P. Birman. An agent-based current differential relay for use with a utility intranet. *IEEE Transactions on Power Delivery*, 17(1):47–53, Jan. 2002.

[48] M. Baran, R. Sreenath, and N. R. Mahajan. Extending EMTDC/PSCAD for simulating agent-based distributed applications. *IEEE Power Engineering Review*, 22(12):52–54, Dec. 2002.

[49] T. Nagata and H. Sasaki. A multi-agent approach to power system restoration. *IEEE Transactions on Power Systems*, 17(2):457–462, May 2002.

[50] H. Ni, G. T. Heydt, and L. Mili. Power system stability agents using robust wide-area control. *IEEE Power Engineering Review*, 22(9):58–58, Sep. 2002.

[51] J. Jung, C-C. Liu, S. L. Tanimoto, and V. Vittal. Adaptation in load shedding under vulnerable operating conditions. *IEEE Transactions on Power Systems*, 17(4):1199–1205, Nov. 2002.

[52] D. P. Buse, P. Sun, Q. H. Wu, and J. Fitch. Agent-based substation automation. *IEEE Power and Energy Magazine*, 1(2):50–55, Mar.–Apr. 2003.

[53] J. A. Hossack, J. Menal, S. D. J. McArthur, and J. R. McDonald. A multiagent architecture for protection engineering diagnostic assistance. *IEEE Transactions on Power Systems*, 18(2):639–647, May 2003.

[54] H. F. Wang, H. Li, and H. Chen. Coordinated secondary voltage control to eliminate voltage violations in power system contingencies. *IEEE Transactions on Power Systems*, 18(2):588–595, May 2003.

[55] S. D. J. McArthur, E. M. Davidson, G. J. W. Dudgeon, and J. R. McDonald. Toward a model integration methodology for advanced applications in power engineering. *IEEE Transactions on Power Systems*, 18(3):1205–1206, Aug. 2003.

[56] V. S. Koritarov. Real-world market representation with agents. *IEEE Power and Energy Magazine*, 2(4):39–46, Jul.–Aug. 2004.

[57] S. D. J. McArthur, S. M. Strachan, and G. Jahn. The design of a multi-agent transformer condition monitoring system. *IEEE Transactions on Power Systems*, 19(4):1845–1852, Nov. 2004.

[58] A. J. Bagnall and G. D. Smith. A multiagent model of the uk market in electricity generation. *IEEE Transactions on Evolutionary Computation*, 9(5):522–536, Oct. 2005.

[59] S. D. J. McArthur, C. D. Booth, J. R. McDonald, and I. T. McFadyen. An agent-based anomaly detection architecture for condition monitoring. *IEEE Transactions on Power Systems*, 20(4):1675–1682, Nov. 2005.

[60] R. Giovanini, K. Hopkinson, D.V. Coury, and J. S. Thorp. A primary and backup cooperative protection system based on wide area agents. *IEEE Transactions on Power Delivery*, 21(3):1222–1230, Jul. 2006.

[61] S. Sheng, K. K. Li, W. L. Chan, Z. Xiangjun, and D. Xianzhong. Agent-based self-healing protection system. *IEEE Transactions on Power Delivery*, 21(2):610–618, Apr. 2006.

[62] S-J. Park and J-T. Lim. Modelling and control of agent-based power protection systems using supervisors. *IEE Proceedings – Control Theory and Applications*, 153(1):92–98, Jan. 2006.

[63] E. M. Davidson, S. D. J. McArthur, J. R. McDonald, T. Cumming, and I. Watt. Applying multi-agent system technology in practice: automated management and analysis of SCADA and digital fault recorder data. *IEEE Transactions on Power Systems*, 21(2):559–567, May 2006.

[64] H. S. Oh and R. J. Thomas. A diffusion-model-based supply-side offer agent. *IEEE Transactions on Power Systems*, 21(4):1729–1735, Nov. 2006.

[65] K. Hopkinson, Xiaoru Wang, R. Giovanini, J. Thorp, K. Birman, and D. Coury. Epochs: a platform for agent-based electric power and communication simulation built from commercial off-the-shelf components. *IEEE Transactions on Power Systems*, 21(2):548–558, May 2006.

[66] A.C. Tellidou and A. G. Bakirtzis. Agent-based analysis of capacity withholding and tacit collusion in electricity markets. *IEEE Transactions on Power Systems*, 22(4):1735–1742, Nov. 2007.

[67] K. Huang, S. K. Srivastava, and D. A. Cartes. Solving the information accumulation problem in mesh structured agent system. *IEEE Transactions on Power Systems*, 22(1):493–495, Feb. 2007.

[68] S. D. J. McArthur, E. M. Davidson, V. M. Catterson, A. L. Dimeas, N. D. Hatziargyriou, F. Ponci, and T. Funabashi. Multi-agent systems for power engineering applications – Part II: Technologies, standards, and tools for building multi-agent systems. *IEEE Transactions on Power Systems*, 22(4):1753–1759, Nov. 2007.

[69] T. Sueyoshi and G. R. Tadiparthi. Agent-based approach to handle business complexity in u.s. wholesale power trading. *IEEE Transactions on Power Systems*, 22(2):532–543, May 2007.

[70] N. Liu, J. Zhang, and W. Liu. A security mechanism of web services-based communication for wind power plants. *IEEE Transactions on Power Delivery*, 23(4):1930–1938, Oct. 2008.

[71] X. Tong, X. Wang, and K. M. Hopkinson. The modeling and verification of peer-to-peer negotiating multiagent colored petri nets for wide-area backup protection. *IEEE Transactions on Power Delivery*, 24(1):61–72, Jan. 2009.

[72] Y. Serizawa, T. Tanaka, H. Yusa, Y. Koda, G. Yamashita, M. Miyabe, S. Katayama, T. Tsuchiya, and K. Omata. Verification of distributed real-time computer network architecture associated with off-the-shelf and dedicated technologies. *IEEE Transactions on Power Delivery*, 24(3):1206–1217, Jul. 2009.

[73] W. Du, Z. Chen, H.F. Wang, and R. Dunn. Feasibility of online collaborative voltage stability control of power systems. *IET Generation, Transmission & Distribution*, 3(2):216–224, Feb. 2009.

[74] N-P. Yu, C-C. Liu, and J. Price. Evaluation of market rules using a multi-agent system method. *IEEE Transactions on Power Systems*, 25(1):470–479, Feb. 2010.

[75] Y. Zhu, S. Song, and D. Wang. Multiagents-based wide area protection with best-effort adaptive strategy. *International Journal of Electrical Power & Energy Systems*, 31(2/3):94–99, 2009.

[76] K. Huang, S. K. Srivastava, D. A. Cartes, and L-H. Sun. Market-based multiagent system for reconfiguration of shipboard power systems. *Electric Power Systems Research*, 79(4):550–556, 2009.

[77] G. Sheng, X. Jiang, D. Duan, and G. Tu. Framework and implementation of secondary voltage regulation strategy based on multi-agent technology. *International Journal of Electrical Power & Energy Systems*, 31(1):67–77, 2009.

[78] C-K. Chang, S-J. Lee, J-K. Park, and H-J. Lee. Application of multi-agent system for overcurrent protection system of industrial power system. In *Proceedings of the International Conference on Intelligent System Applications to Power Systems (ISAP)*, Budapest, Hungary, Jun. 2001.

[79] T. Kato, H. Kanamori, Y. Suzuoki, and T. Funabashi. Multi-agent based control and protection of power distributed system – protection scheme with simplified information utilization. In *Proceedings of the International Conference on Intelligent System Applications to Power Systems (ISAP)*, Arlington, Virginia, USA, Nov. 2005.

[80] A. L. Dimeas and N. D. Hatziargyriou. A mas architecture for microgrids control. In *Proceedings of the International Conference on Intelligent System Applications to Power Systems (ISAP)*, Arlington, Virginia, USA, Nov. 2005.

[81] I. H. Lim, S. J. Lee, M. S. Choi, and P. Crossley. Multi-agent system-based protection coordination of distribution feeders. In *Proceedings of the International Conference on Intelligent System Applications to Power Systems (ISAP)*, Niigata, Japan, Nov. 2007.

[82] E. M. Davidson and S. D. J. McArthur. Exploiting multi-agent system technology within an autonomous regional active network management system. In *Proceedings of the International Conference on Intelligent System Applications to Power Systems (ISAP)*, Niigata, Japan, Nov. 2007.

[83] T. Tsuji, T. Goda, K. Ikeda, and S. Tange. Autonomous decentralized voltage profile control of distribution network considering time-delay. In *Proceedings of the International Conference on Intelligent System Applications to Power Systems (ISAP)*, Niigata, Japan, Nov. 2007.

[84] Y. L. Lo, C. H. Wang, and C. N. Lu. A multi-agent based service restoration in distribution network with distributed generations. In *Proceedings of the International Conference on Intelligent System Applications to Power Systems (ISAP)*, Curitiba, Brazil, Nov. 2009.

[85] A. L. Dimeas and N. D. Hatziargyriou. Control agents for real microgrids. In *Proceedings of the International Conference on Intelligent System Applications to Power Systems (ISAP)*, Curitiba, Brazil, Nov. 2009.

[86] Y-T. Pan and M-S. Tsai. Development a bdi-based intelligent agent architecture for distribution systems restoration planning. In *Proceedings of the International Conference on Intelligent System Applications to Power Systems (ISAP)*, Curitiba, Brazil, Nov. 2009.

[87] I-H. Lim, M-S. Choi, S-J. Lee, and T. W. Kim. Intelligent distributed restoration by multi-agent system concept in DAS. In *Proceedings of the International Conference on Intelligent System Applications to Power Systems (ISAP)*, Curitiba, Brazil, Nov. 2009.

[88] E. M. Davidson, M. J. Dolan, S. D. J. McArthur, and G. W. Ault. The use of constraint programming for the autonomous management of power flows. In *Proceedings of the International*

Conference on Intelligent System Applications to Power Systems (ISAP), Curitiba, Brazil, Nov. 2009.

[89] W-Y. Yu, V-W. Soo, M-S. Tsai, and Y-B. Peng. Coordinating a society of switch agents for power distribution service restoration in a smart grid. In *Proceedings of the International Conference on Intelligent System Applications to Power Systems (ISAP)*, Athens, Greece, Sep. 2011.

[90] A. L. Dimeas and N. D. Hatziargyriou. Operation of a multiagent system for microgrid control. *IEEE Transactions on Power Systems*, 20(3):1447–1455, 2005.

[91] M. M. Nordman and M. Lehtonen. Distributed agent-based state estimation for electrical distribution networks. *IEEE Transactions on Power Systems*, 20(2):652 –658, May 2005.

[92] M. M Nordman and M. Lehtonen. An agent concept for managing electrical distribution networks. *IEEE Transactions on Power Delivery*, 20(2):696–703, Apr. 2005.

[93] I. S. Baxevanos and D. P. Labridis. Implementing multiagent systems technology for power distribution network control and protection management. *IEEE Transactions on Power Delivery*, 22(1):433–443, Jan. 2007.

[94] I. S. Baxevanos and D. P. Labridis. Software agents situated in primary distribution networks: A cooperative system for fault and power restoration management. *IEEE Transactions on Power Delivery*, 22(4):2378–2385, Oct. 2007.

[95] M. E. Baran and I. M. El-Markabi. A multiagent-based dispatching scheme for distributed generators for voltage support on distribution feeders. *IEEE Transactions on Power Systems*, 22(1):52–59, Feb. 2007.

[96] J. M. Solanki, S. Khushalani, and N. N. Schulz. A multi-agent solution to distribution systems restoration. *IEEE Transactions on Power Systems*, 22(3):1026–1034, Aug. 2007.

[97] N. Perera, A. D. Rajapakse, and T. E. Buchholzer. Isolation of faults in distribution networks with distributed generators. *IEEE Transactions on Power Delivery*, 23(4):2347 –2355, Oct. 2008.

[98] T. Xu, N. S. Wade, E. M. Davidson, P. C. Taylor, S. D. J. McArthur, and W. G. Garlick. Case-based reasoning for coordinated voltage control on distribution networks. *Electric Power Systems Research*, 81(12):2088–2098, 2011.

[99] A. S. Bouhouras, G. T. Andreou, and D. P. Labridis. Feasibility study of the implementation of a.i. automation techniques in modern power distribution networks. *Electric Power Systems Research*, 80(5):495–505, 2010.

[100] T. Logenthiran, D. Srinivasan, and A. M. Khambadkone. Multi-agent system for energy resource scheduling of integrated microgrids in a distributed system. *Electric Power Systems Research*, 81(1):138–148, 2011.

[101] S. M. Brahma and A. A. Girgis. Development of adaptive protection scheme for distribution systems with high penetration of distributed generation. In *Proceedings of the IEEE PES General Meeting*, Toronto, Canada, Jul. 2003.

[102] S. Chouhan, H. Wan, H. J. Lai, A. Feliachi, and M. A. Choudhry. Intelligent reconfiguration of smart distribution network using multi-agent technology. In *Proceedings of the IEEE PES General Meeting*, Calgary, Canada, Jul. 2009.

[103] L. A. Zadeh. Fuzzy sets as a basis for theory of possibility. *Fuzzy Sets and Systems*, 1:3–28, 1978.

[104] H. Wan, K. K. Li, and K. P. Wong. An adaptive multiagent approach to protection relay coordination with distributed generators in industrial power distribution system. *IEEE Transactions on Industry Applications*, 46(5):2118–2124, Sep.–Oct. 2010.

[105] László Z. Varga, N. R. Jennings, and D. Cockburn. Integrating intelligent systems into a cooperating community for electricity distribution management. *Expert Systems with Applications*, 7(4):563–579, Oct.–Dec. 1994.

[106] N. R. Jennings, E. H. Mamdani, J. M. Corera, I. Laresgoiti, F. Perriolat, P. Skarek, and L. Z. Varga. Using archon to develop real-world DAI applications. *IEEE Expert*, 11(6):64–70, Dec. 1996.

[107] J. A. Hossack, S. D. J. McArthur, and J. R. McDonald. Integrating intelligent protection analysis tools using multi-agent technologies. In *Proceedings of the International Conference on Intelligent System Applications to Power Systems (ISAP)*, Lemnos, Greece, Aug.–Sep. 2003.

[108] F. M. T. Brazier, B. M. Dunin-Keplicz, N. R. Jennings, and J. Treur. DESIRE: Modeling multi-agent systems in a compositional formal framework. *International Journal of Cooperative Information Systems*, 6(1):69–94, 1997.

[109] J. C. Collis, D. T. Ndumu, H. D. Nwana, and L. C. Lee. ZEUS agent building tool-kit. *BT Technology Journal*, 16(3):60–68, Jul. 1998.

[110] E. E. Mangina, S. D. J. McArthur, and J. R. McDonald. COMMAS: Condition monitoring multi-agent system. *Autonomous Agents and Multi-Agent Systems*, 4(3):279–282, Sep. 2001.

[111] T. Nagata, Y. Tao, H. Sasaki, and H. Fujita. A multiagent approach to distribution system restoration. In *Proceedings of the IEEE PES General Meeting*, volume 2, Toronto, Canada, Jul. 2003.

[112] T. Nagata, Y. Tao, K. Kimura, H. Sasaki, and H. Fujita. A multi-agent approach to distribution system restoration. In *Proceedings of the IEEE Midwest Symposium on Circuits and Systems*, volume 2, Hiroshima, Japan, Jul. 2004.

[113] J. M. Solanki and N. N. Schulz. Multi-agent system for islanded operation of distribution systems. In *Proceedings of the IEEE PES Power Systems Conference and Exposition*, pages 1735–1740, Atlanta, USA, Nov. 2006.

[114] G. Zhabelova and V. Vyatkin. Multiagent smart grid automation architecture based on IEC 61850/61499 intelligent logical nodes. *IEEE Transactions on Industrial Electronics*, 59(5):2351–2362, May 2012.

[115] IEC 61850. Communication networks and systems in substations. International Electrotechnical Commitee, 2003.

[116] IEC 61499. Function blocks. International Electrotechnical Commitee, 2005.

[117] MATLAB. http://www.mathworks.com/.

[118] D. M. Staszesky. Use of virtual agents to effect intelligent distribution automation. In *Proceedings of the IEEE PES General Meeting*, Montreal, Canada, Oct. 2006.

[119] S&C Electric Company. IntelliTeam SG significantly reduces outage impact at Midwest city. http://www.sandc.com/news/?p=1832, Nov. 2011.

[120] F. van Overbeeke. Active networks: Distribtuion networks facilitating integration of distributed generation. In *Proceedings of the International Symposium on Distributed Generation: Power System and Market Aspects*, Stockholm, Sweden, Oct. 2002.

[121] P.H. Nguyen, W. L. Kling, G. Georgiadis, M. Papatriantafilou, L. A. Tuan, and L. Bertling. Distributed routing algorithms to manage power flow in agent-based active distribution network. In *Proceedings of the IEEE PES Innovative Smart Grid Technologies Conference (ISGT) Europe*, Stockholm, Sweden, Oct. 2010.

[122] C. Rehtanz. *Autonomous Systems and Intelligent Agents in Power System Control and Operation*. Springer, New York, USA, Sep. 2003.

[123] P. H. Nguyen, W. L. Kling, and P. F. Ribeiro. Smart power router: A flexible agent-based converter interface in active distribution networks. *IEEE Transactions on Smart Grid*, 2(3):487–495, Sep. 2011.

[124] ABB. Station computer COM615. `http://www.abb.com/product/db0003db004281/c12573e700 330419c1257188002facd2.aspx`, Apr. 2012.

[125] E. M. Davidson, M. J. Dolan, G. W. Ault, and S. D. J. McArthur. AuRA-NMS: An autonomous regional active network management system for edf energy and sp energy networks. In *Proceedings of the IEEE PES General Meeting*, Minneapolis, USA, Jul. 2010.

[126] P. Taylor, T. Xu, S. McArthur, G. Ault, E. Davidson, M. Dolan, C. Yuen, M. Larsson, D. Botting, D. Roberts, and P. Lang. Integrating voltage control and power flow management in aura-nms. In *Proceedings of the IET-CIRED SmartGrids for Distribution CIRED Seminar*, Frankfurt, Germany, Aug. 2008.

[127] J. K. Kok, C. J. Warmer, and I. G. Kamphuis. Powermatcher: Multiagent control in the electricity infrastructure. In *Proceedings of the International Joint Conference on Autonomous Agents and Multiagent Systems*, Utrecht, Netherlands, Jul. 2005.

[128] B. Roossie. Field-test upscaling of multi-agent coordination in the electricity grid. In *Proceedings of the International Conference and Exibition on Electricity Distribution*, Frankfurt, Germany, Jun. 2009.

[129] A. Dimeas and N. Hatziargyriou. A multiagent system for microgrids. In *Proceedings of the IEEE PES General Meeting*, Denver, USA, Jun. 2004.

[130] A. L. Dimeas and N. D. Hatziargyriou. Agent based control for microgrids. In *Proceedings of the IEEE PES General Meeting*, Tampa, USA, Jun. 2007.

[131] S. J. Chatzivasiliadis, N. D. Hatziargyriou, and A. L. Dimeas. Development of an agent based intelligent control system for microgrids. In *Proceedings of the IEEE PES General Meeting – Conversion and Delivery of Electrical Energy in the 21st Century*, Pittsburgh, USA, Jul. 2008.

[132] A. L. Dimeas, S. I. Hatzivasiliadis, and N. D. Hatziargyriou. Control agents for enabling customer-driven microgrids. In *Proceedings of the IEEE PES General Meeting*, Calgary, Canada, Jul. 2009.

[133] A. G. Madureira, J. C. Pereira, N. J. Gil, J. A. Peças Lopes, G. N. Korres, and N. D. Hatziargyriou. Advanced control and management functionalities for multi-microgrids. *European Transactions on Electrical Power*, 21(2):1159–1177, Mar. 2011.

[134] D. A. Cohen. Gridagents: Intelligent agent applications for integration of distributed energy resources within distribution systems. In *Proceedings of the IEEE PES General Meeting – Conversion and Delivery of Electrical Energy in the 21st Century*, Pittsburgh, USA, Jul. 2008.

[135] A. A. Aquino-Lugo, R. Klump, and T. J. Overbye. A control framework for the smart grid for voltage support using agent-based technologies. *IEEE Transactions on Smart Grid*, 2(1):173–180, Mar. 2011.

[136] M. E. Elkhatib, R. El-Shatshat, and M. M. A. Salama. Novel coordinated voltage control for smart distribution networks with dg. *IEEE Transactions on Smart Grid*, 2(4):598–605, Dec. 2011.

[137] H. E. Z. Farag, E. F. El-Saadany, and R. Seethapathy. A two ways communication-based distributed control for voltage regulation in smart distribution feeders. *IEEE Transactions on Smart Grid*, 3(1):271–281, Mar. 2012.

[138] A. Pokahr, L. Braubach, and W. Lamersdorf. Jadex: Implementing a BDI-infrastructure for JADE agents. *EXP – in search of innovation (Special Issue on JADE)*, 3(3):76–85, Sep. 2003.